Table of Contents

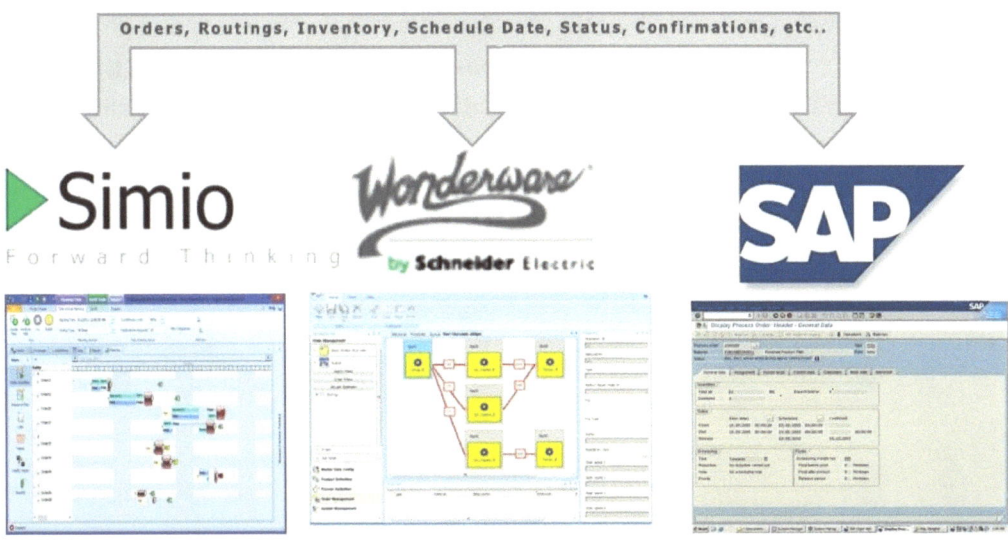

Overview

Welcome to the world of simulation. Simulation provides a unique way to examine the future and make intelligent decisions based on what you learn. While simulation technology has been around for decades, it is still rapidly evolving. Advances in object-oriented approaches provide rapid modeling and the flexibility to model complex systems that could not be modeled just a few years ago. And integrated 3D animation makes the creation of compelling 3D visualizations easy, and this in turn helps assure more robust, understandable models and better communication with stakeholders.

Simio extends that base technology to provide **a unique solution** for planning and scheduling. Simio provides Wonderware MES with Finite Capacity Scheduling and Advanced Planning and Scheduling functionality. Simio's patented approach to planning and scheduling is first to automate comparing a deterministic schedule from MRP with a variation adjusted schedule to let you **effectively deal with breakdowns, unplanned events and material shortages**. You can then re-schedule resources to acceptable levels of risk. Data is brought into memory for **fast execution**. While Simio takes full advantage of your existing data to quickly and automatically build you scheduling model, that base model can also be enhanced as needed to **accurately capture complexity unique to your system**.

The team of architects behind Simio have been leaders in simulation since the early 1980's, playing key roles in the design and development of four previous market-leading products. Simio is the result of this team applying their collective 180 years of simulation experience and using the very latest in technology and development techniques to create a new generation of simulation problem solving capability.

ERP Integration

The Simio Production Scheduling model is typically integrated with the existing ERP and MES/SCADA system to provide the data on the current status of the system and the actual jobs to be processed through the system. This data is provided to Simio in the form of relational data sets that are imported and held in memory by Simio for fast execution. These data sets typically contain data such as a list of jobs to be processed, a bill of material for each job, job routings including setup and processing times, purchased material, etc., along with the status of all jobs in the system at the start of the planning period. Simio does not have a fixed data schema – the data sets used by Simio Scheduling Edition are configured as needed in the Simio Enterprise Edition to match the form of the external data. The flow of jobs as defined by the data sets is then simulated by Simio Scheduling Edition using the custom Simio model to generate a detailed schedule. A schematic of a typical implementation is shown below:

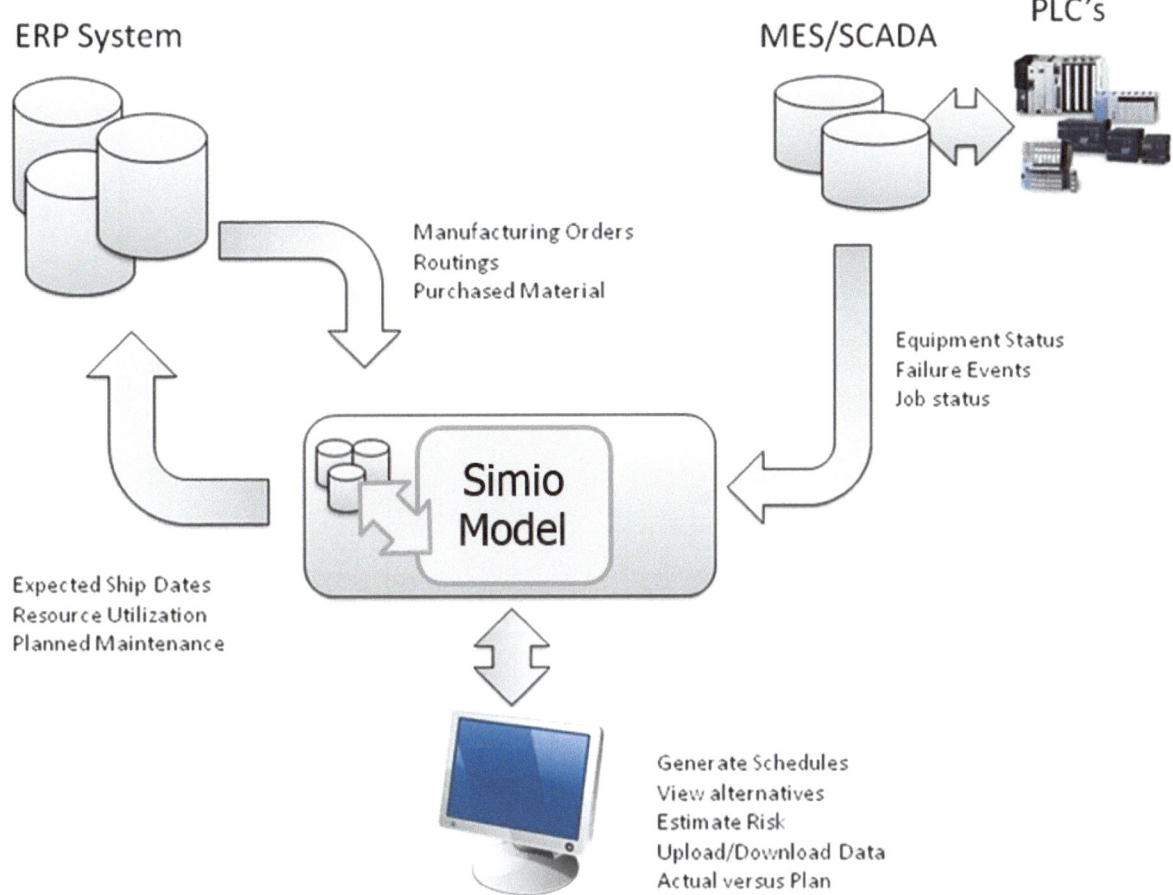

Integration between Simio Production Scheduling and Wonderware MES

This interface has been built to integrate data between Simio Production Scheduling and Wonderware MES to enable real-time event based scheduling. Utilizing Simio's simulation-based scheduling capabilities, a model can be developed to depict the resource and material constraints within a manufacturing system. Then, interfacing Simio to the Wonderware MES enables the scheduling model to be run near real-time. This mean as events happen on the factory floor, the production schedule can be re-run quickly to re-route orders around possible bottlenecks.

The integration employs pulling of data from tables and views of the MES Table to get data into the model. The Wondeware MES data bind uses the Wonderware MES API to pull this data from the Wonderware MES data model. Inventory, work orders, routings, bill-of-materials, order status and machine status are all captured through this inbound integration to pull the data for creating the production schedule.

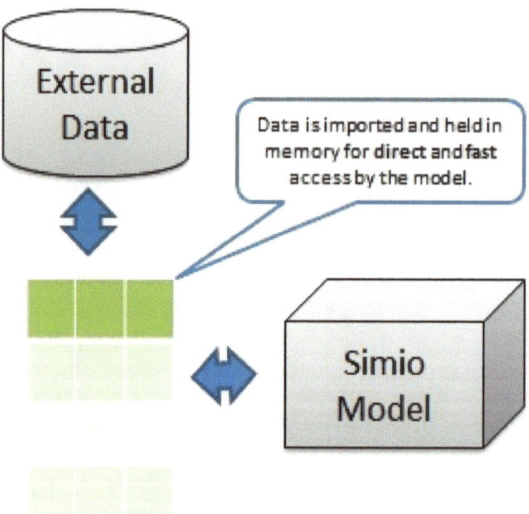

Data is imported and held in
memory for direct and fast
access by the model.

Once the schedule is generated, the scheduled is updated in the Wonderware MES using the Wonderware MES API. The Export Schedule to MES add-in has been developed to communicate with the Wonderware MES API. This add-in is used to update schedule dates, scheduled resource and status on MES jobs.

Interface Components

There are a number of components that are used for this integration. These components can be downloaded using the following URL:

http://www.simio.com/downloads/public/software/WonderwareMES.zip

Unzip the file. The contents are as follows:

- DashboardAndReport – Contains dashboard and table report files. The dashboards have an "xml" suffix and the table reports have an "repx" suffix.
- DatabaseScripts - SQL views, function and stored procedures used to pull the MES data.
- DemoModels – There are 2 completed scheduling models. One model shows the integration with WonderwareMES and the other shows the integration with InBatch.
- Documents – This folder contains 2 documents. The first this SimioAPI-UsingTheSimioBackendDll.pdf that shows how to run Simio without the GUI. The MiddlewareExtensibility_Jan2013.pdf is a reference on how to trigger evens from the WondewareMES Middleware.
- InBatchXMLStylesheets – This folder contains a sample InBatch recipe in BatchML format as well as 4 XML Stylesheets used to transform the BatchML into the format needed to import the data into the Simio tables.

- RunSimioSchedule – The folder contains the RunSimioSchedule executable and supporting files to run Simio Schedule using a Windows Service.
- RunSimioScheduleCode – The folder contains the visual studio project used to generated the RunSimioSchedule executable.
- TestHooks – The folder contains the visual studio project that produces a file when configured to run via the WonderwareMES middleware.
- Wonderware Data Bind / Add-Ins / Automatic Scheduling code – The Wonderware Data Bind and Add-Ins are installed with Simio.   The code for the Wonderware Data Bind and Add-In is provided so it can be extended for a project. In addition, the code and .NET Assemblies used to automatic schedule are provided.
- Simio Models – Simulation models that are various phases of development.   These models show the component (physically layout, the scheduling logic, database schema and dashboard/reports) for creating an integrated production schedule between Simio and Wonderware MES.
- WondwareMES – the folder contains the Wonderware Data Bind and Add-Ins are installed with Simio.   The code for the Wonderware Data Bind and Add-In is provided so it can be extended for a project.
- Wonderware MES 2014 SP1 database backup – A Microsoft SQL Server 2014 database backup is provided to recreate the demo if someone has Wonderware MES 2014 SP1.

The DatabaseScript need to be run on the WonderwareMES database to setup the integration. The other folders are optional depending on the project. If needed, they will be referenced in other sections of this document.

Name	Date modified	Type
DashboardsAndReports	2/3/2017 6:41 PM	File folder
DatabaseScripts	2/3/2017 6:46 PM	File folder
DemoModels	2/3/2017 6:47 PM	File folder
Documents	2/3/2017 6:41 PM	File folder
InBatchXMLStylesheets	2/3/2017 7:26 PM	File folder
RunSimioSchedule	2/3/2017 7:00 PM	File folder
RunSimioScheduleCode	2/3/2017 6:41 PM	File folder
TestHooks	2/3/2017 6:41 PM	File folder
WonderwareMESCode	2/3/2017 6:42 PM	File folder
WonderwareMESMSSQL2014Backup	2/3/2017 6:53 PM	File folder

Install Database Object

The database views, functions and stored procedures are used to map the data between the Wonderware MES database and the Simio table. A database object is defined in each SQL script provided. The scripts are to be run on the same database used by Wonderware MES.

Install the following scripts:

- fn_Simio_String_Fomat.sql – Function to convert data to Simio String Format. Any object names that start with a number will have 'N_' placed in front. Also, any object names that have a dash ('-') will be converted to an underscore ('_').
- Simio_Ents_View.sql – Lists of all entities that are defined as 'can_sched_jobs' or 'can_run_jobs'.
- sp_Simio_Ents –The stored procedure calls the Simio_Ents_View. I also add 2 additional object called WOCreate and WOShip. These objects typically not found in the WonderwareMES database. They are used to create and dispose orders in the model.
- Simio_Items_View.sql – Lists all items (materials) from WonderwareMES.
- Simio_Item_Inv_View.sql –List all inventory. This view shows all items and their current inventory by lot.
- Simio_Work_Orders_View.sql – Lists all work orders in MES from the wo table
- sp_Simio_Update_WO_Dates –The stored procedure re-sets the wo release_time and req_finished_time fields to the current data and then returns the results of the Simio_Work_Orders_View. This stored procedure is used mainly for demo purposes to keep the demo up to date.
- Simio_Jobs_View.sql – List all jobs in MES
- sp_Simio_Jobs –The stored procedure calls the Simio_Jobs_View. I also add an additional operation for WOShip is added for each workorder. This operation typically not found in the WonderwareMES database.
- Simio_Job_Boms_View.sql – Lists all job based bill-of-materials.
- Simio_Util_Log_View – List any current down time events in MES. NOTE: Only needed to import downtime events into Simio. This view only works in MES 2014.
- Simio_Processes_View.sql – List all processes in MES. NOTE: This view is only needed if work orders exist in MES that do not have jobs. (e.g. The process_id in the workorder table will be populated when importing from the Simio_Work_Orders_View).
- Simio_Opers_View.sql – List all operations in MES. NOTE: This view is only needed if work orders exist in MES that do not have jobs.
- sp_Simio_Jobs –The stored procedure calls the Simio_Oper_View. I also add an additional operation for WOShip is added for each material. This operation typically not found in the WonderwareMES database.
- Simio_Item_Boms_View.sql – Lists all item based bill-of-materials. NOTE: This view is only needed if work orders exist in MES that do not have jobs and job_boms.

New Simio Model

Create a new Simio model.   Then, select the Data tab.   Select the "Add Wonderware MES Lists and Tables" add-in.

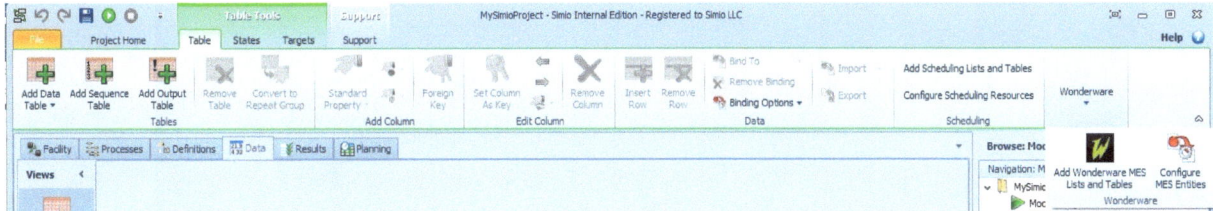

When prompted, select "MES Order Based" as the Routing Type and press OK.

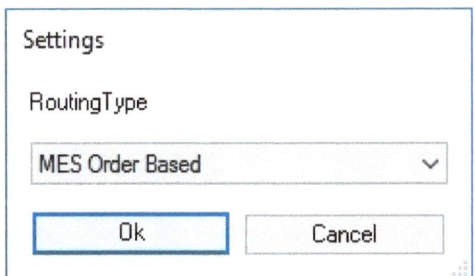

This will add the data tables, lists and configure the model to work with WonderwareMES.

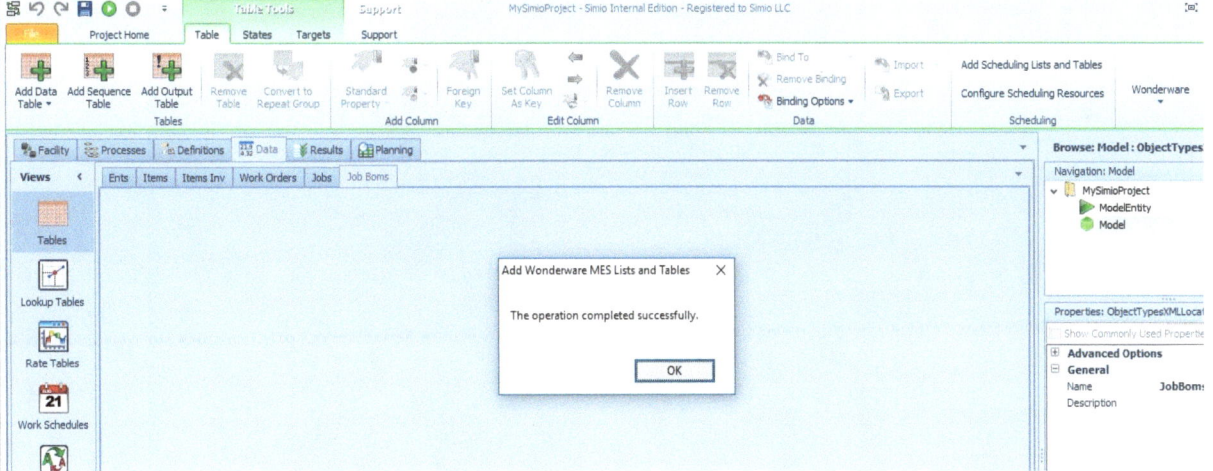

Import Entities

In this step, we will import the entities (referred to as resources in Simio) from Wonderware MES into Simo. Select the 'Data' tab and then select the 'Ents' table. Then, select the 'Bind To' option and specify 'Wonderware MES'.

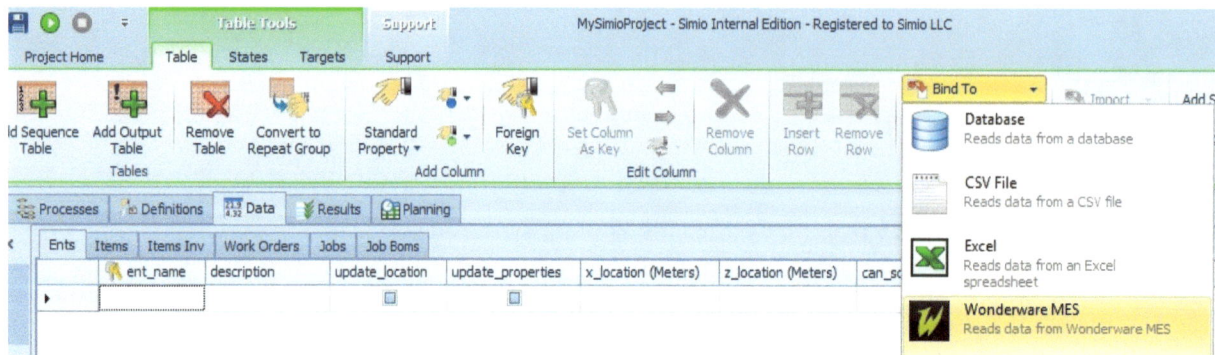

Select the "Is Stored Procedure" check box and enter "sp_Simio_Ents". Then press 'Preview'.

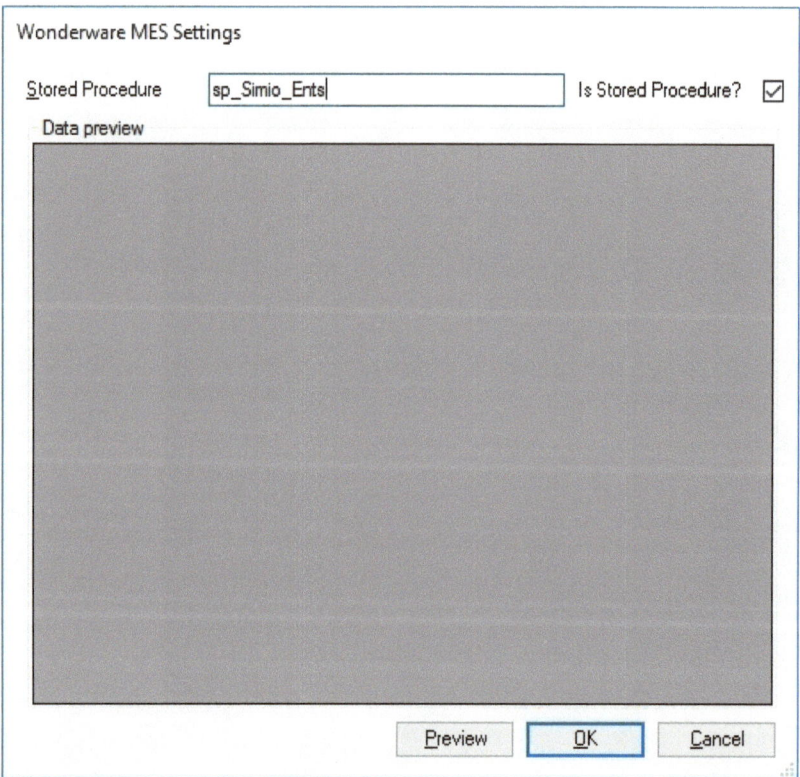

If prompted, enter Wonderware MES User Name and Password.

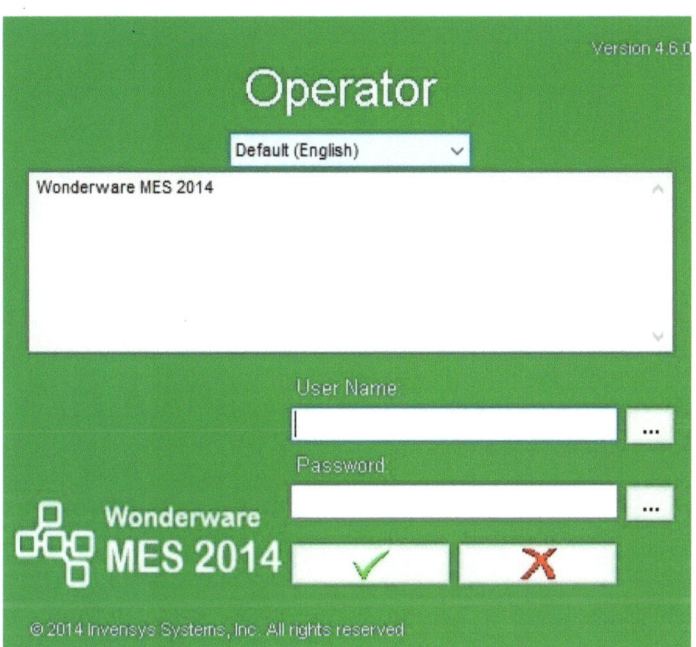

Once preview is displayed, press 'OK'

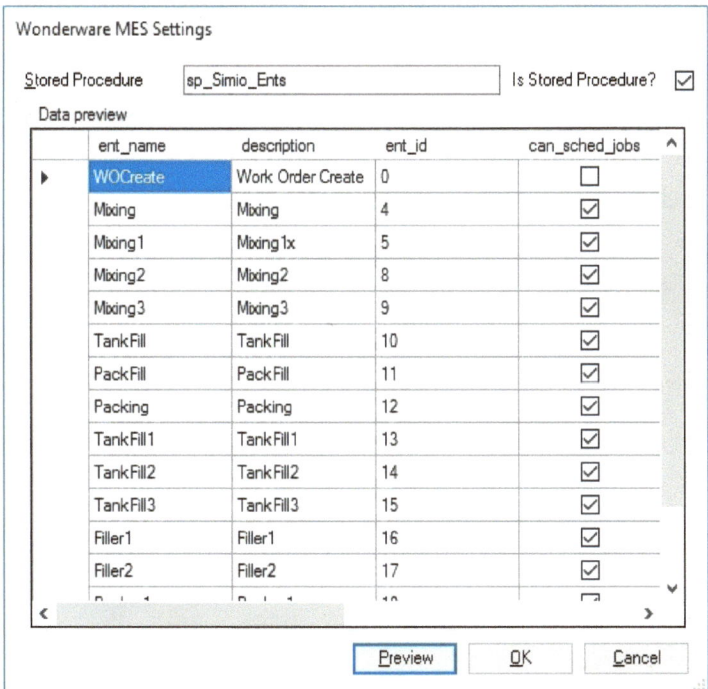

Press the import button. This will import the records from the 'sp_Simio_Ents' stored procedure into the Ents table.

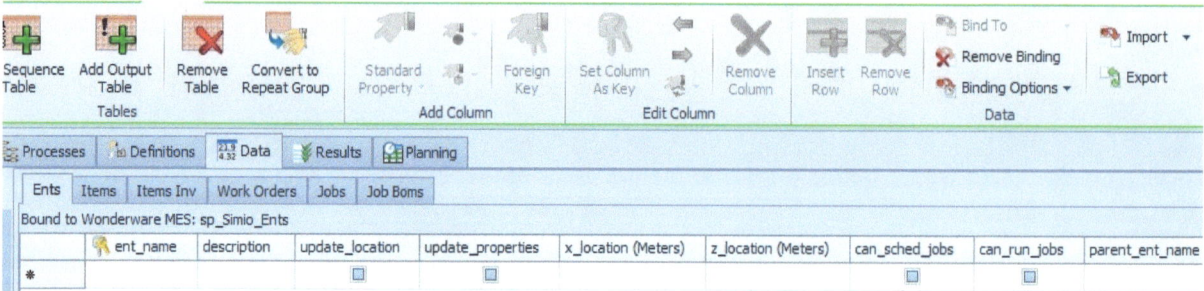

You will see a number of errors at the bottom. The errors exist because the objects have not yet been created in the model. The missing objects will be added in the next step.

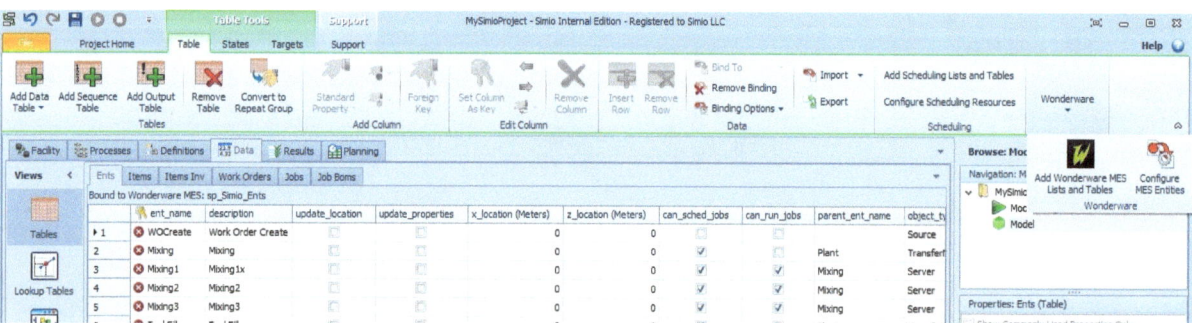

Select the 'Configure MES Entities' add-in from the Wonderware group. This add-in will add a Wonderware MES object for each row in the Ents table. It will also create groupings (node lists) of homogeneous resources. When prompted, select "MES Order Based" as the Routing Type and press OK.

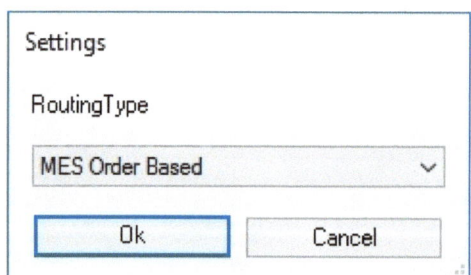

Select the 'Facility' tab. You will find that all the new objects added based on the sequence of the rows in the Ents table.

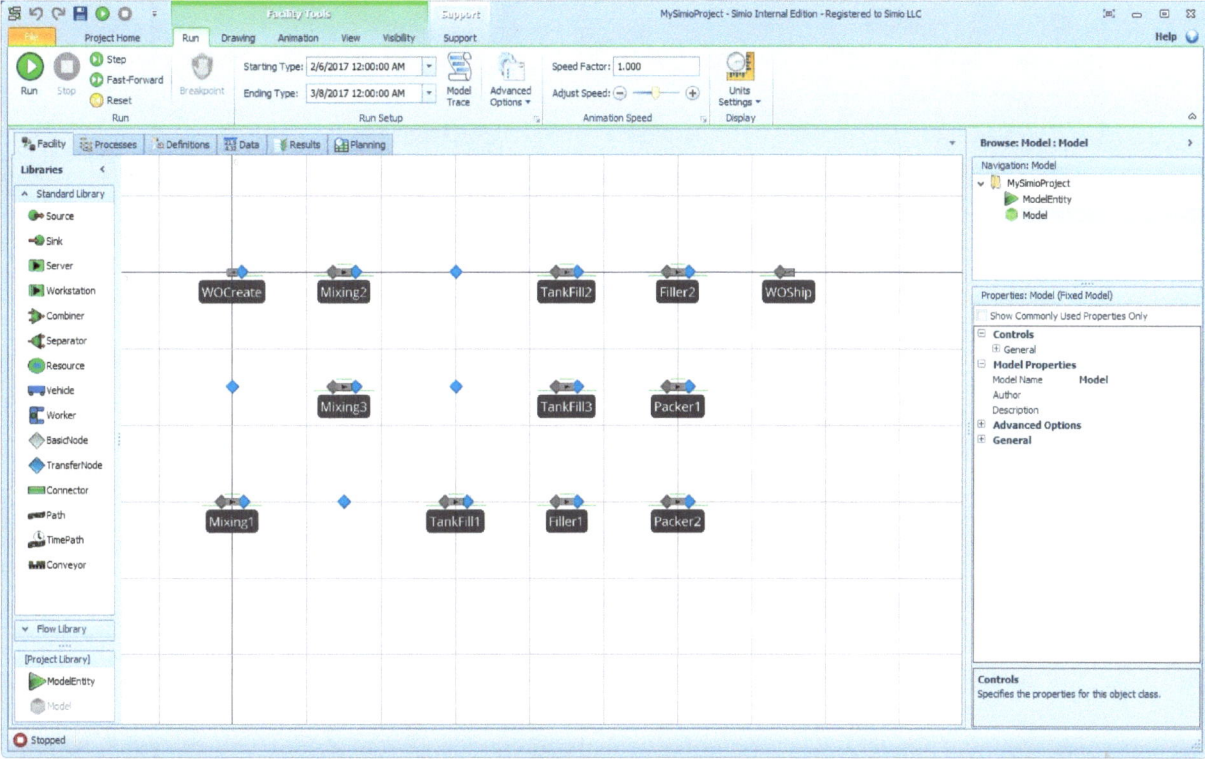

Drag and drop each object to match the Facility layout. Once done, your Facility view is ready to use. It is good to save your model at this time. You can choose File...Save As to choose a different name for the model.

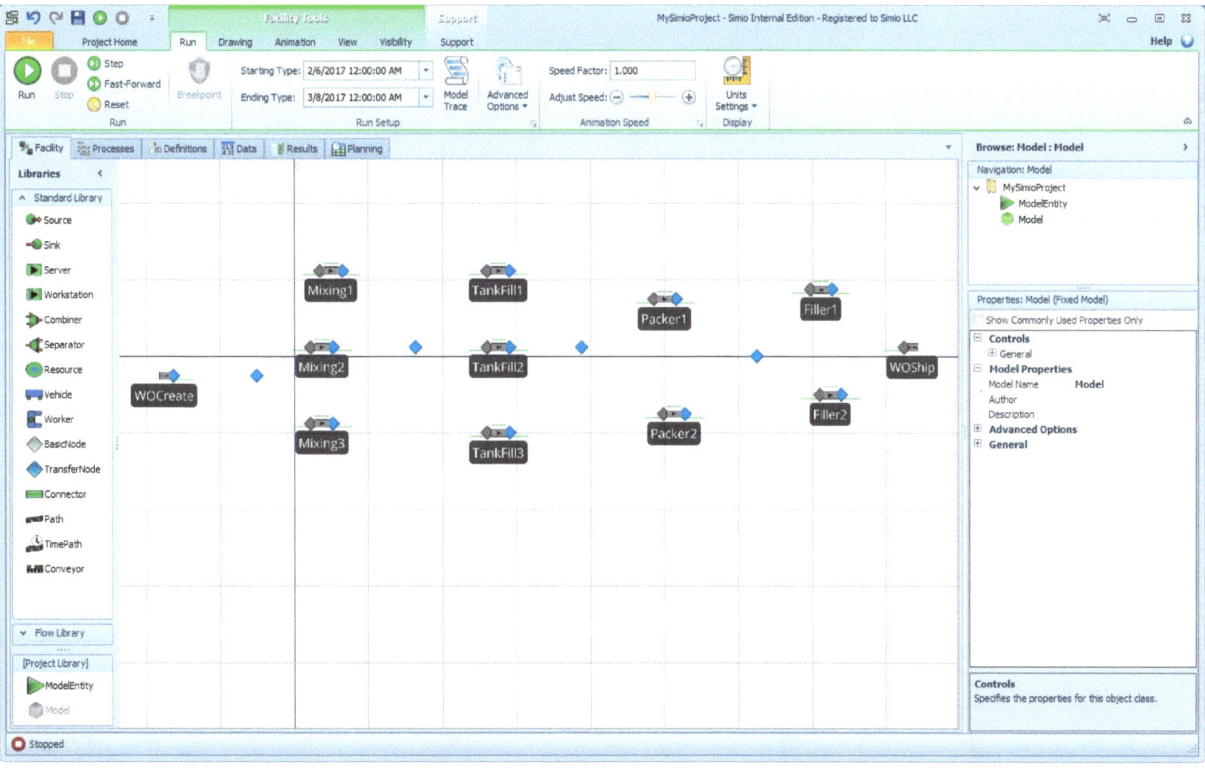

Import Product Definition and Order Management Tables

Follow the same database binding steps described in the Import Entities section to import the rest of the data.

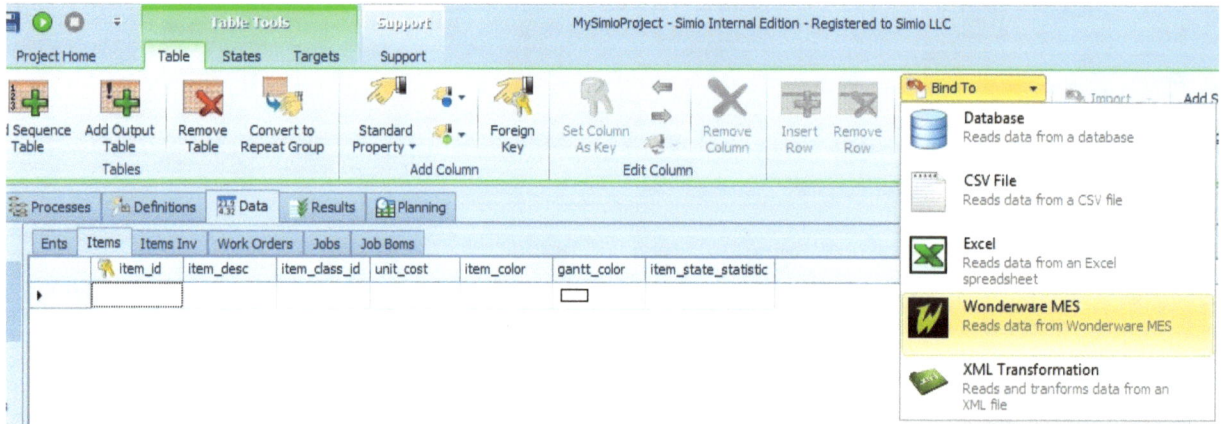

It is suggested that you import the views in the following sequence. If this sequence is followed, no errors should occur.

- Item - Simio_Items_View
- Items_Inv – Simio_Item_Inv_VIew
- Simio_Work_Orders_View
- Jobs - sp_Simio_Jobs (Select "Is Stored Procedure" option).
- Job Boms - Simio_Job_Boms_View

After binding each table, press the Import option to import the data into the model.

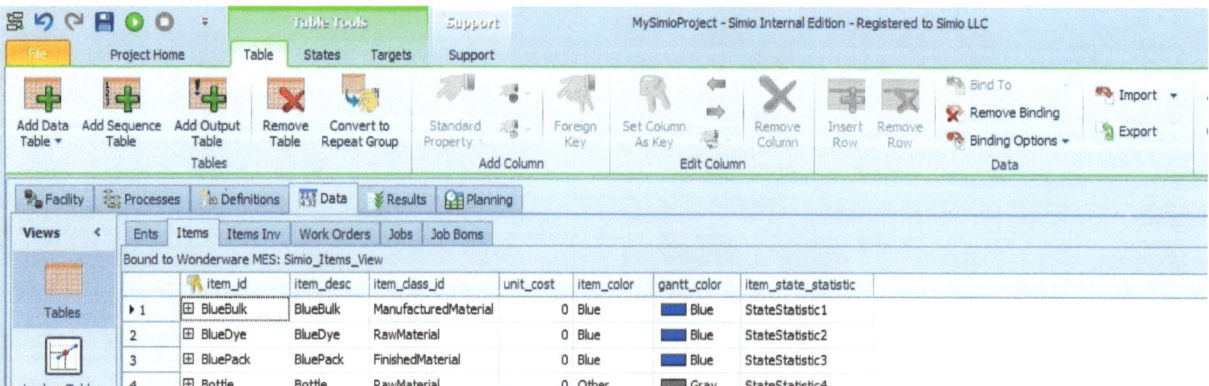

Once completed, there should be no error in the model.

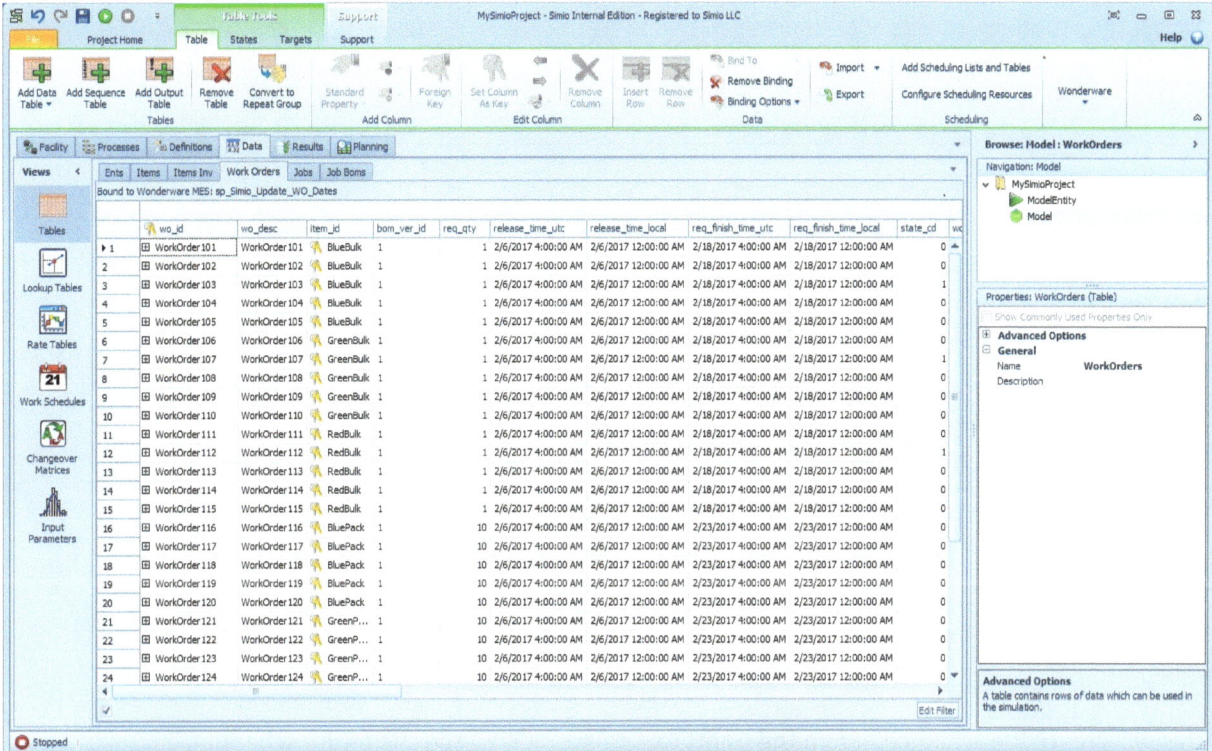

Generating the Schedule

Select the 'Planning' tab. Once selected, the 'Resource Plan" should be populated with the
Wonderware MES Entities. From here, the schedule can be generated. By default, the work orders will
be scheduled based on priority. To create the schedule, select "Create Plan".

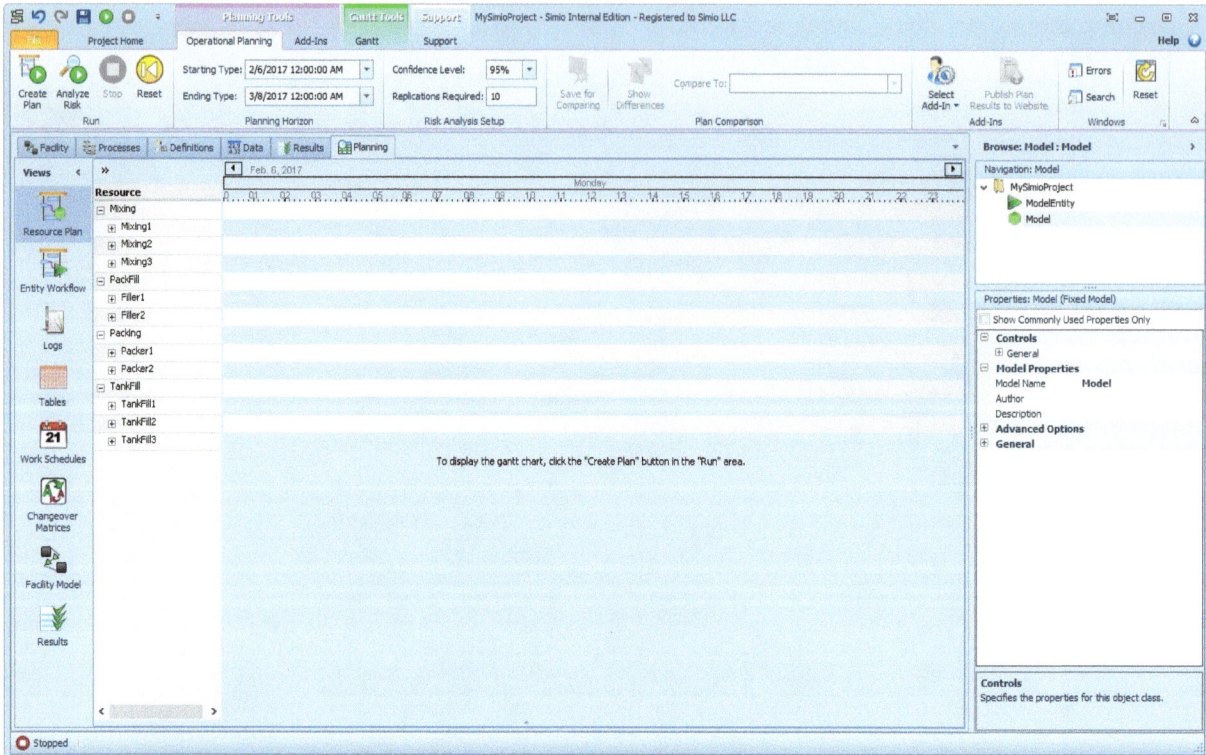

This will create the schedule. From the Gantt tab, select "All" to see the entire schedule.

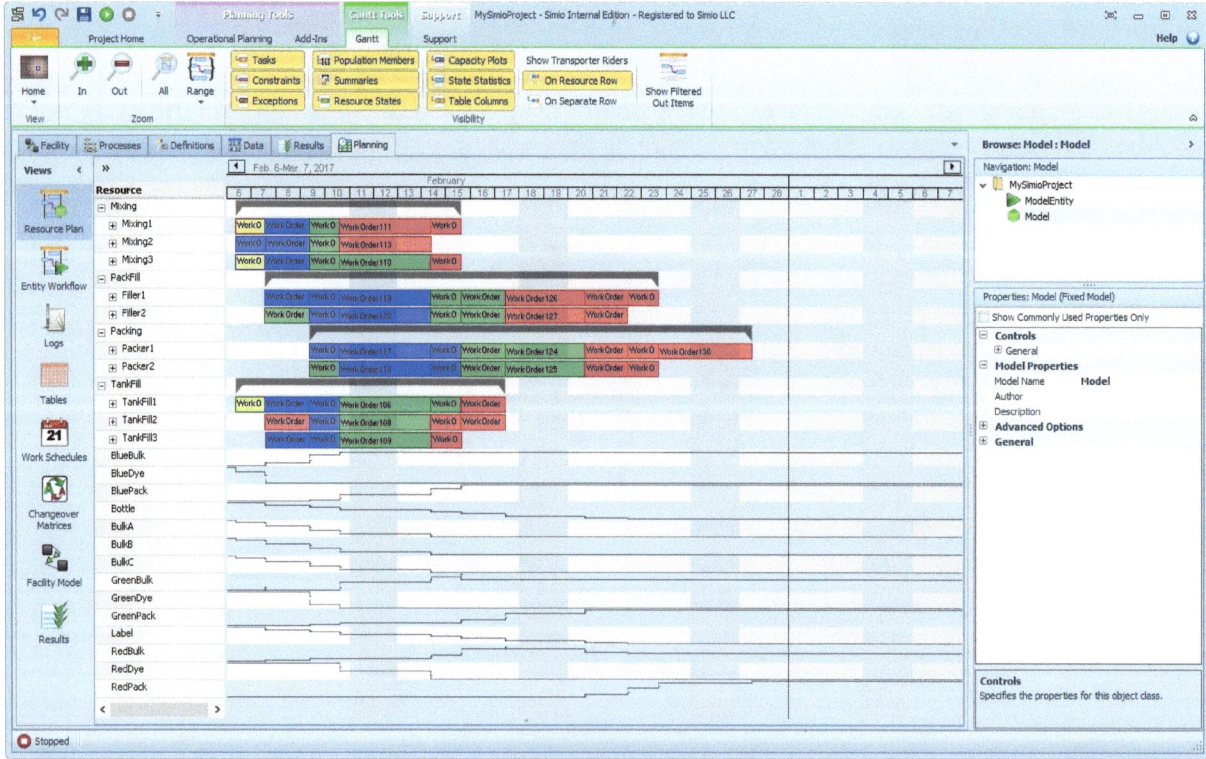

Exporting Results

To save schedule back to Wonderware MES, select Wonderware - Export Schedule to MES. This will save the schedule start, schedule end, schedule resource and set the status of the first job to "Ready"

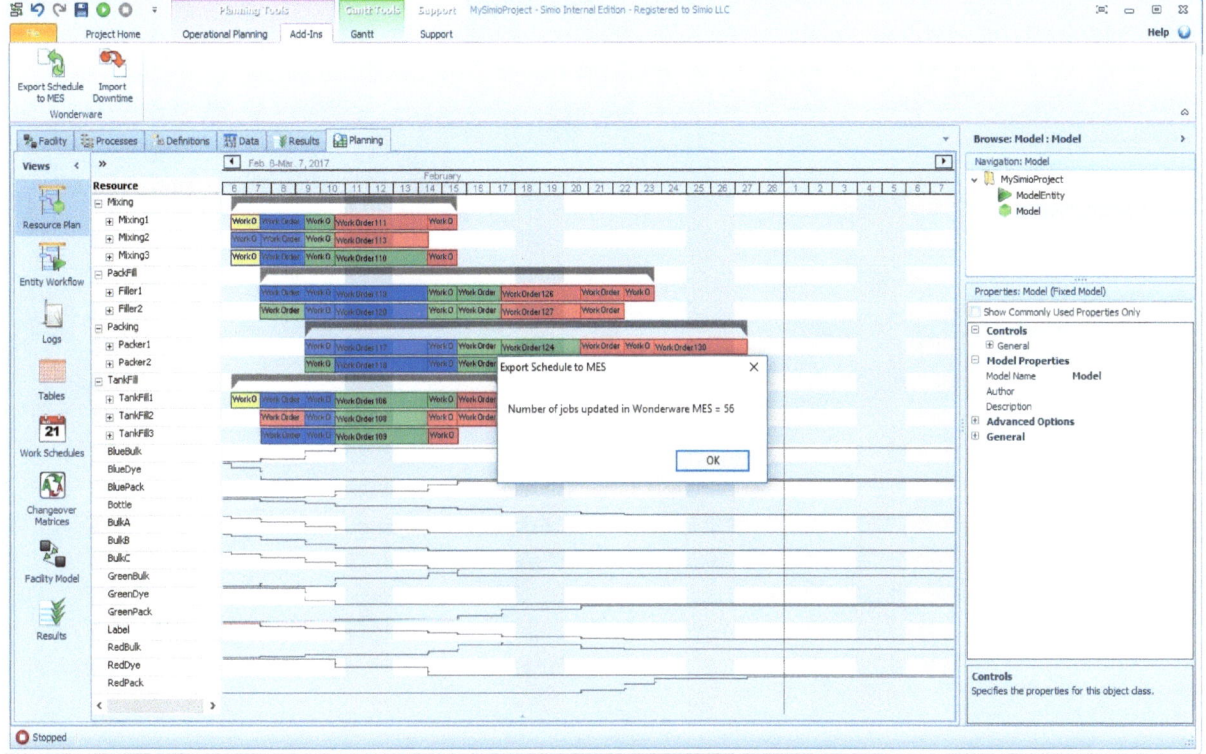

In Wonderware MES, you will see that the jobs have been updated with the Simio scheduling information.

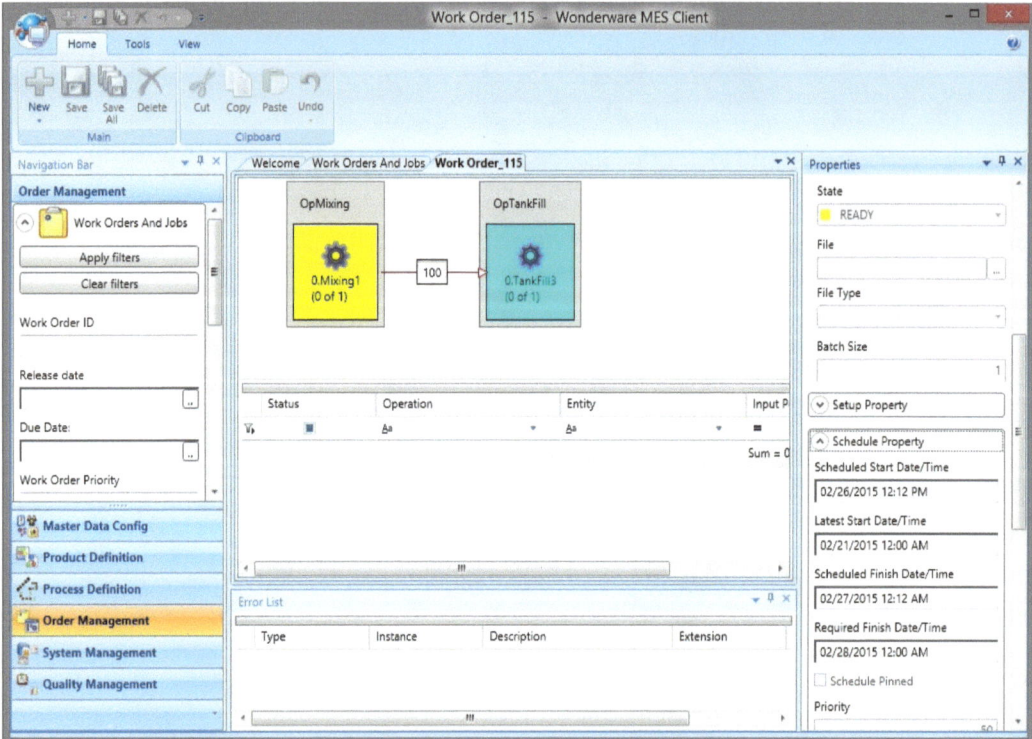

Import Downtime

Add downtime on an entity in Wonderware MES Portal

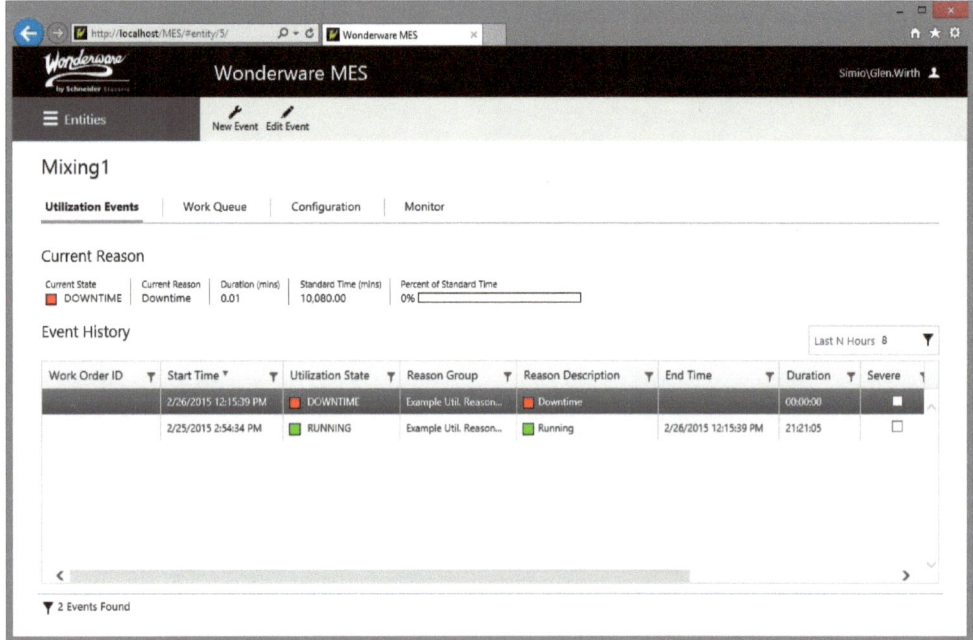

Within Simio, select Wonderware - Import Downtime

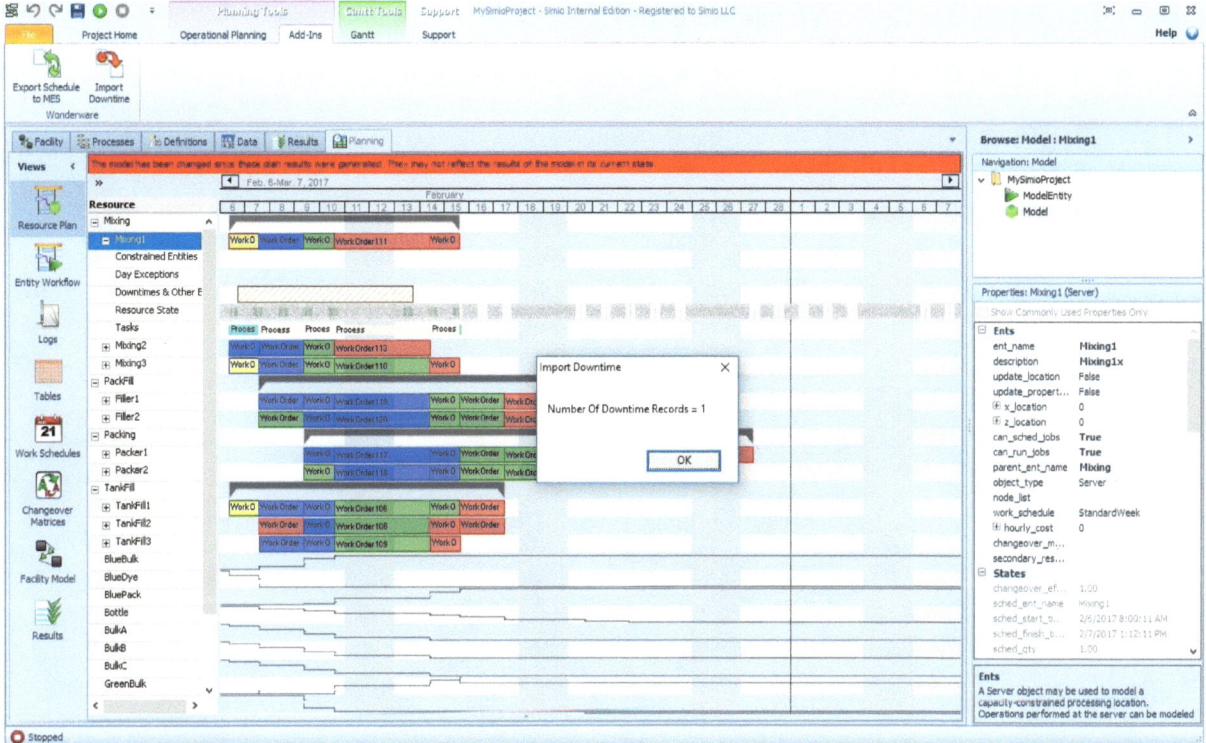

Re-plan around downtime. You will see that jobs have been shifted from Mixer1 to the other mixer resources.

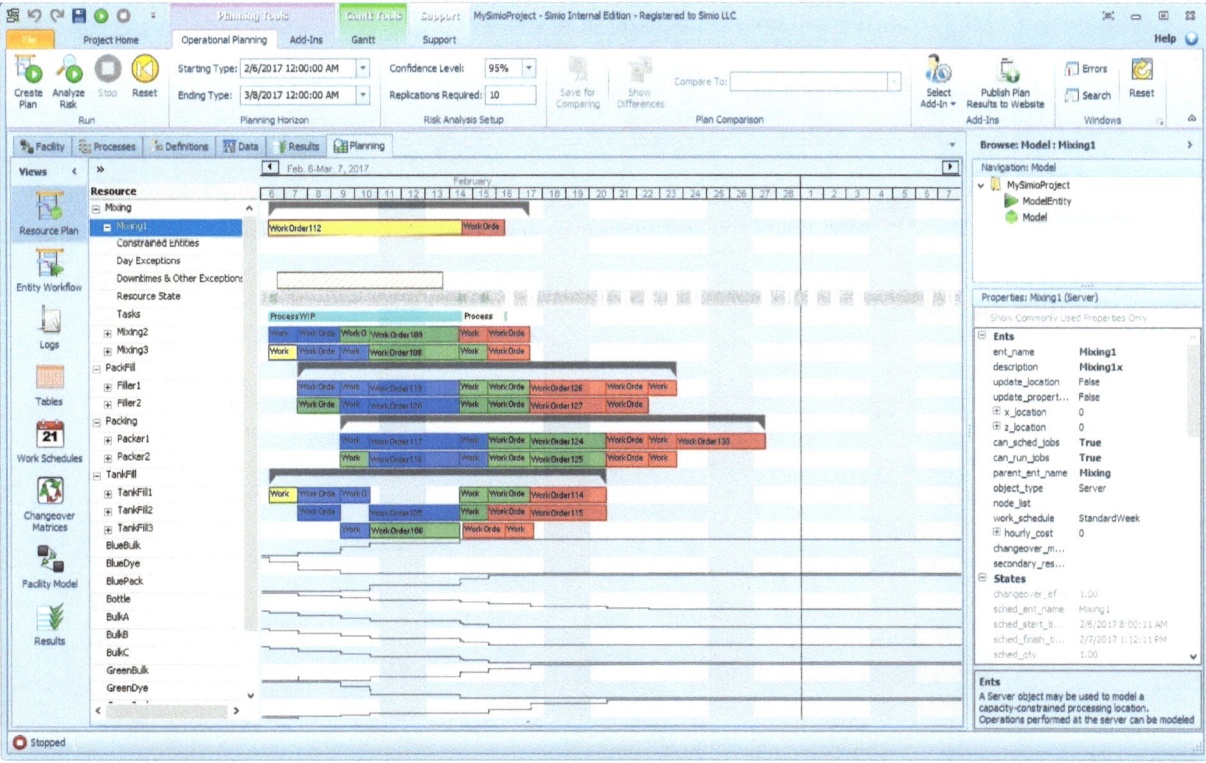

Export updated schedule back to Wonderware MES

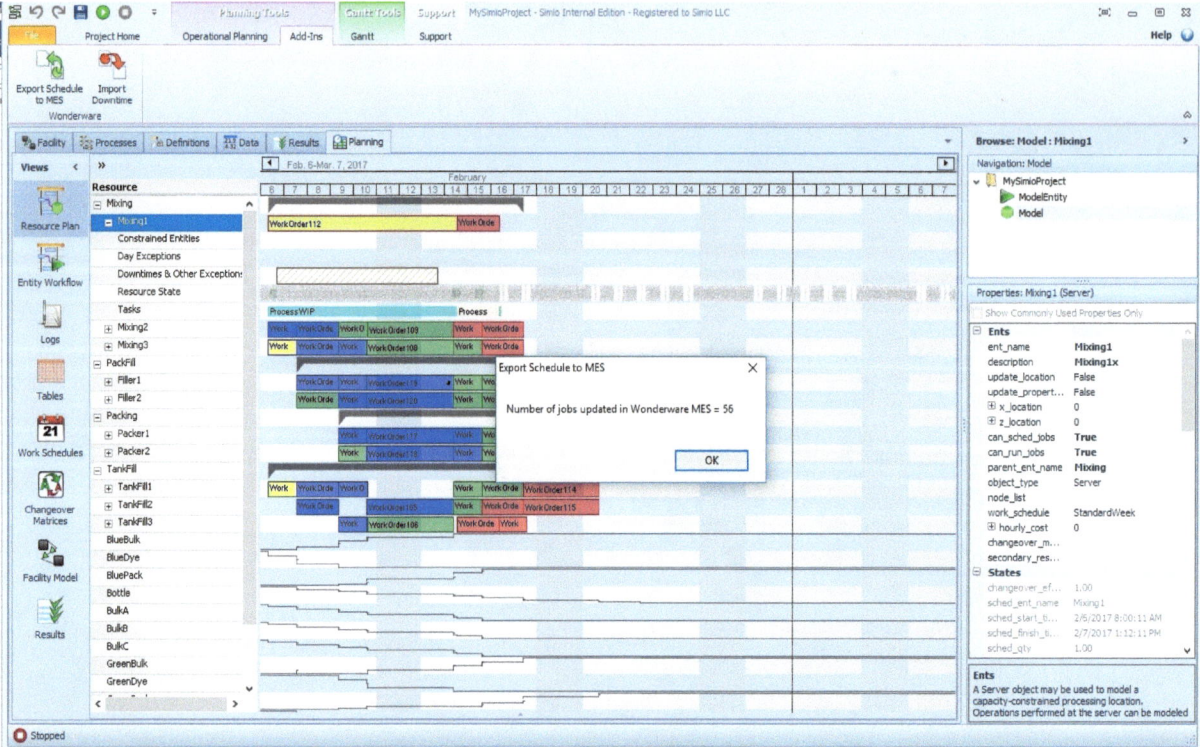

The Wonderware MES schedule will now reflect the downtime event.

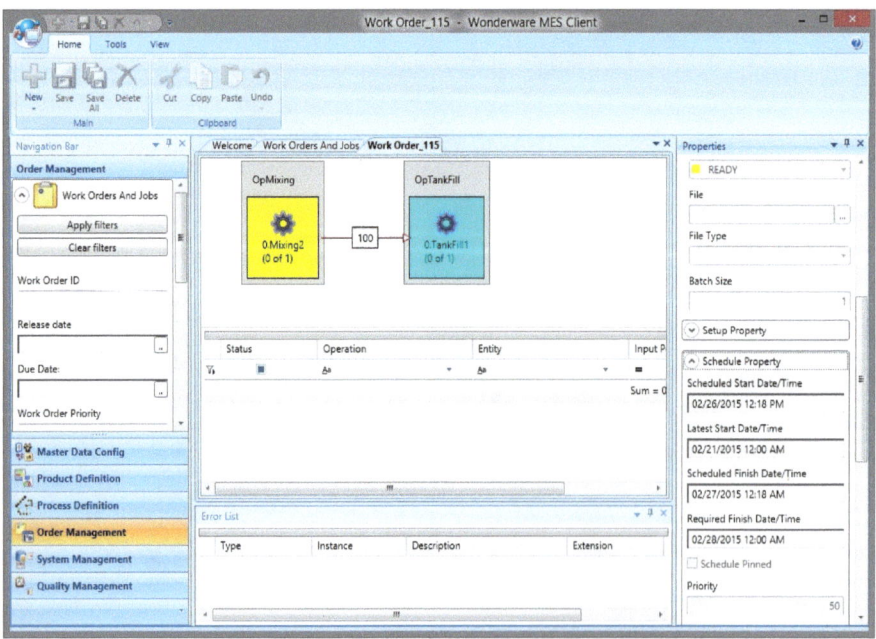

Appendix A – Automatic Reschedule

To automate reschedule, the Wonderware MES Middleware Extensibility Hooks and the Simio Backend DLL are used. The Wonderware MES Middleware Extensibility Hooks are used to capture the events in Wonderware MES. Once the events are captured, the SImio Backended DLL will see the event and automatically re-run the schedule and send the results back to MES.

To setup the Wonderware MES Middleware Extensibility Hooks, the 'Custom Assembly That Is Not in the GAC' example that start on page 26 of the MiddlewareExtensibility_Jan2013.pdf guide was referenced. The example was modified slightly to change the name of the output file and to capture any exceptions. The method used is :

```csharp
public void TestHooksMethod(string xmlSource)
{
    // Compose a string that consists of three lines.
    string lines = string.Format("DateTime: {0}, XMLSource: {1}", DateTime.Now.ToString(), xmlSource);
    // Write the string to a file.
    try
    {
        using (System.IO.StreamWriter file = new System.IO.StreamWriter("C:\\Temp\\Event.txt", true))
        {
            file.WriteLine(lines);
        }

    }
    // Catch Exception
    catch (Exception ex)
    {
        using (System.IO.StreamWriter file = new System.IO.StreamWriter("c:\\Temp\\Error.txt", true))
        {
            file.WriteLine(ex.Message);
        }
    }
}
```

This assembly was compiled using the same structure defined in the 'Custom Assembly That Is Not in the GAC' example.

(C:\Temp\TestHooks\TestHooks\bin\Debug\TestHooks.dll;TestHooksNamespace.TestHooksClass;TestHooksMethod)

Once Within the Middleware Configuration Editor, a post hook is defined to call the assembly every time a new downtime is entered or updated 'SP_U_EXEC_NEWREASON' . When the new downtime is entered / updated, the Event.txt is added / updated.

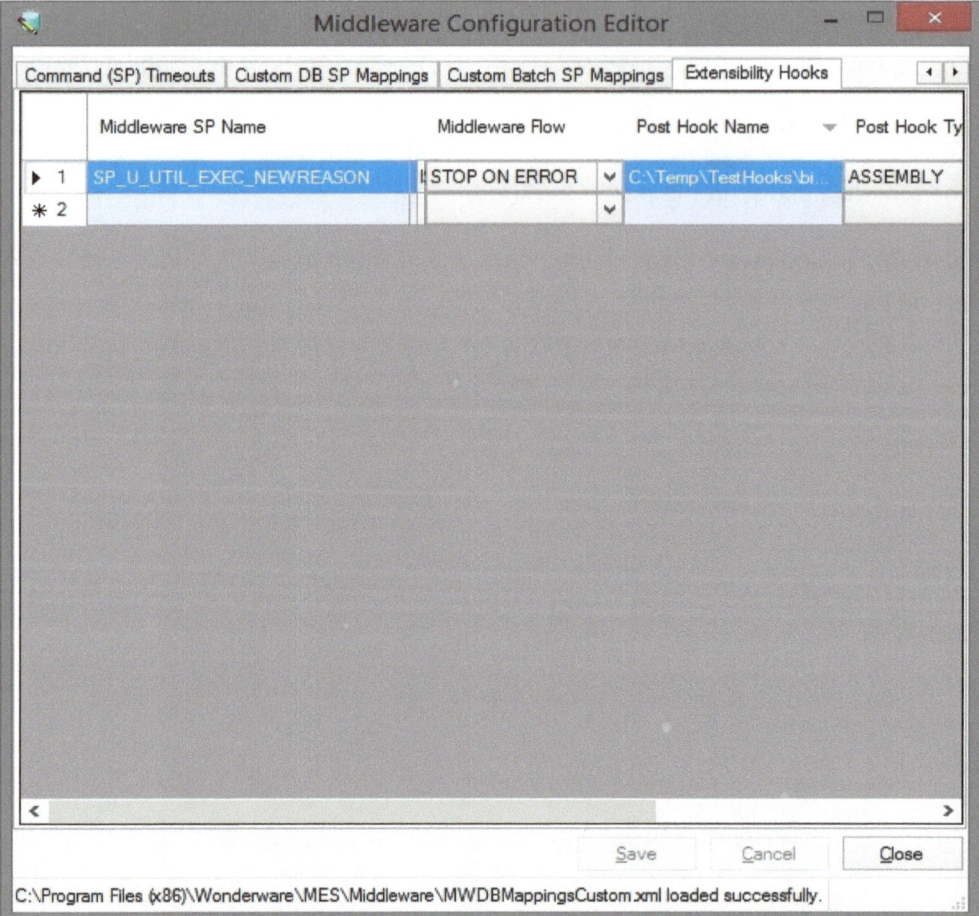

Once configured, the middleware server needs to be restarted before this event will be captured.

Copy the RunSimioSchedule folder from the WonderwareMES.zip into the 'C:\Temp' folder. Modify the RunSimioSchedule.exe.config. Specify the location of your Simio model in the SimioModel setting(e.g. C:\\Temp\\BaseWonderwareMES.spfx). If you followed the instructions in this documentation verbatim, the EventFile and StatusFile settings do not need to be changed.

Next, register the RunSimioSchedule.exe as a windows service. Open a command prompt as an administrator (e.g. right click and select 'Run as Administrator').

Navigate to 'C:\Temp\RunSimioSchedule' From the command prompt, enter command 'InstallUtil RunSimioSchedule.exe'

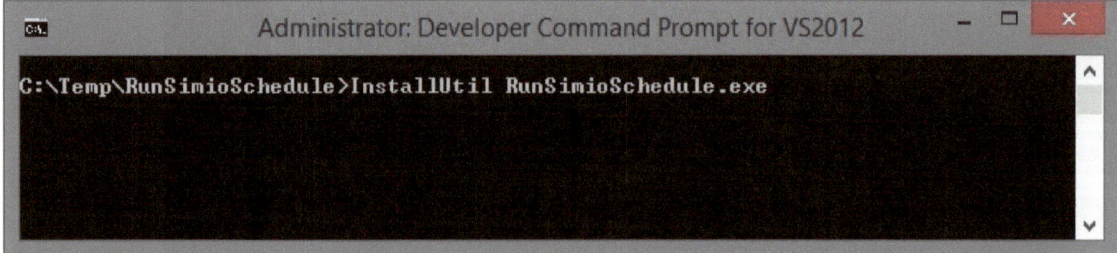

This will install the RunSimioSchedule as a windows service.

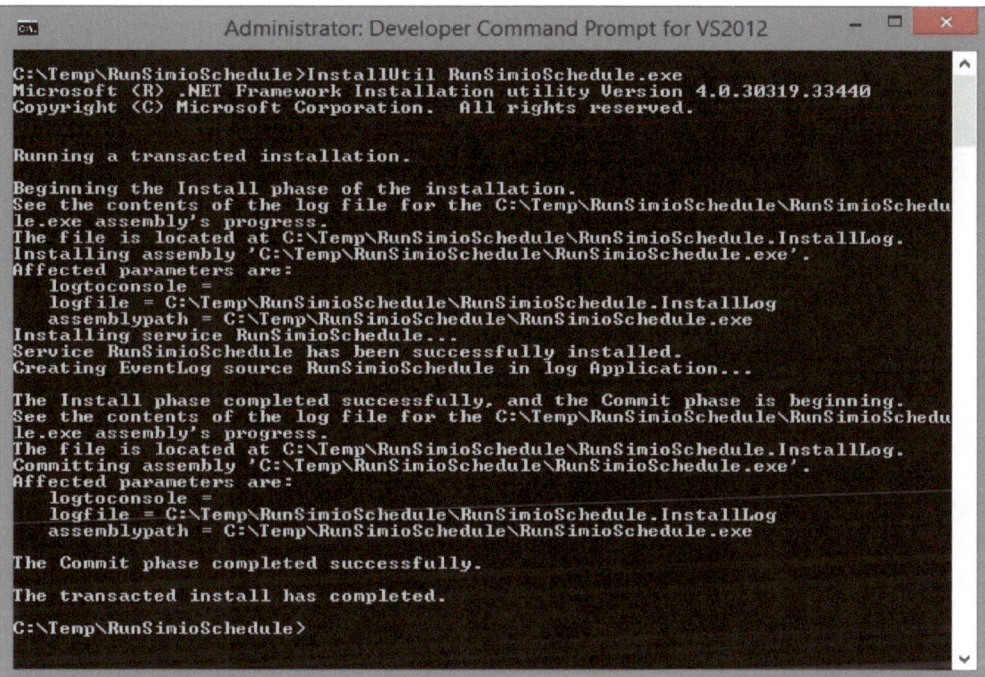

Next, start the service.   Enter 'net start RunSimioService' to start the service.

To test the automatic generation of the schedule, check the work queue to verify that the current dispatch list on an entity.

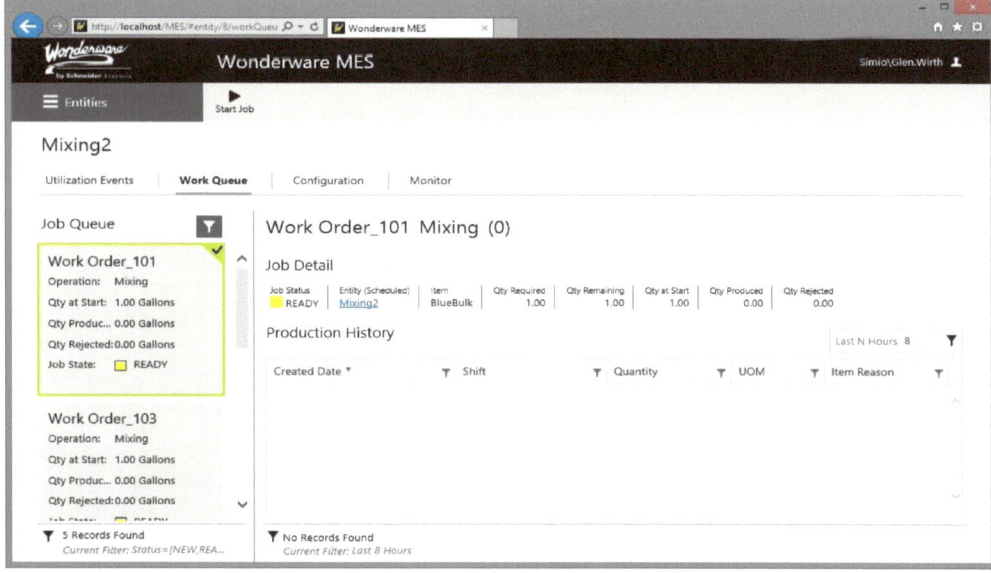

Enter a downtime on the entity….

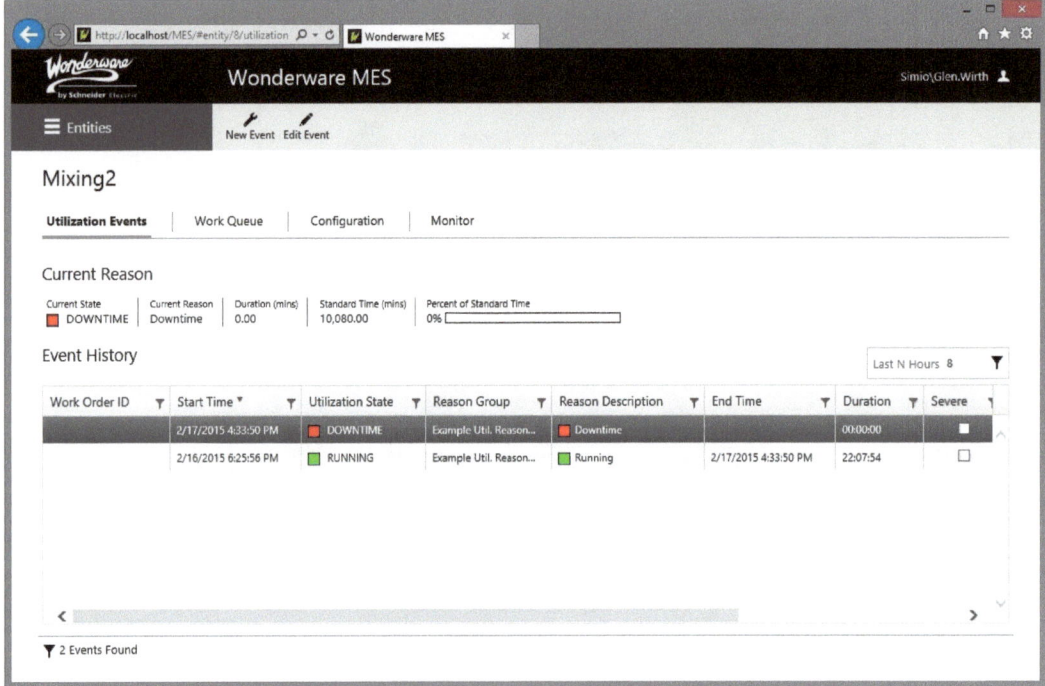

Verify that the schedule has changed.

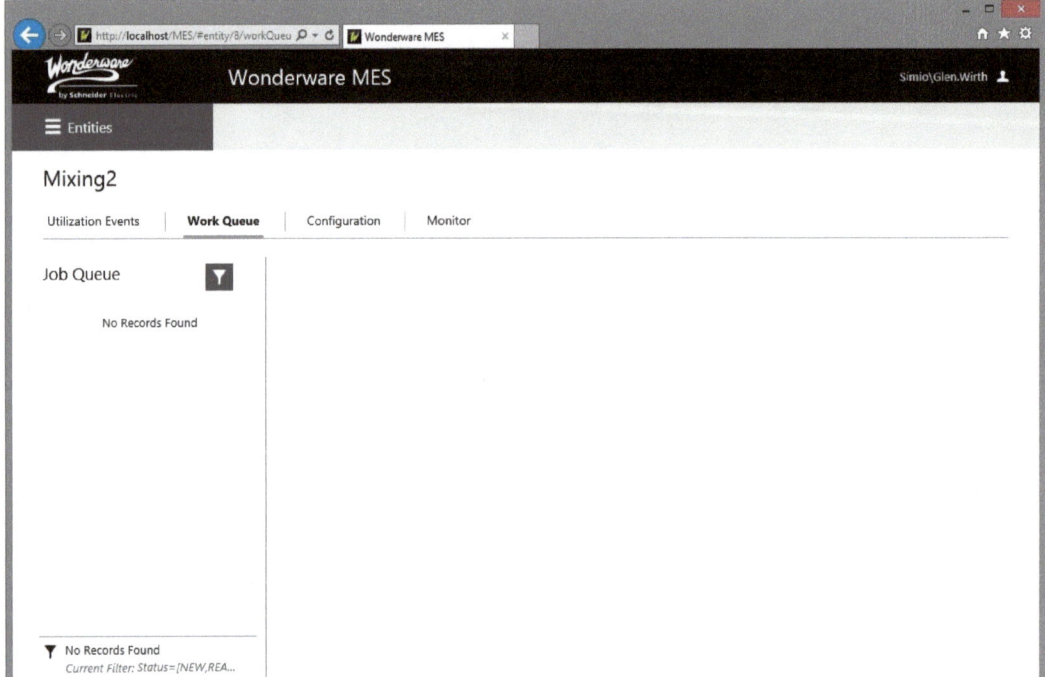

Remove the downtime on the entity

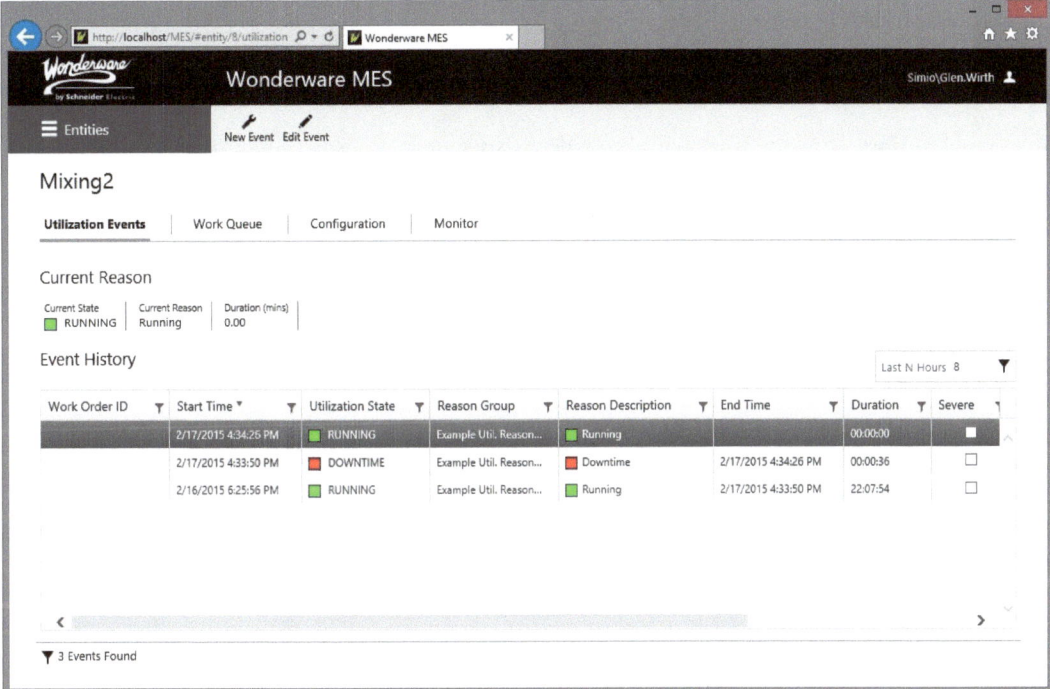

Verify that the entity has been rescheduled.

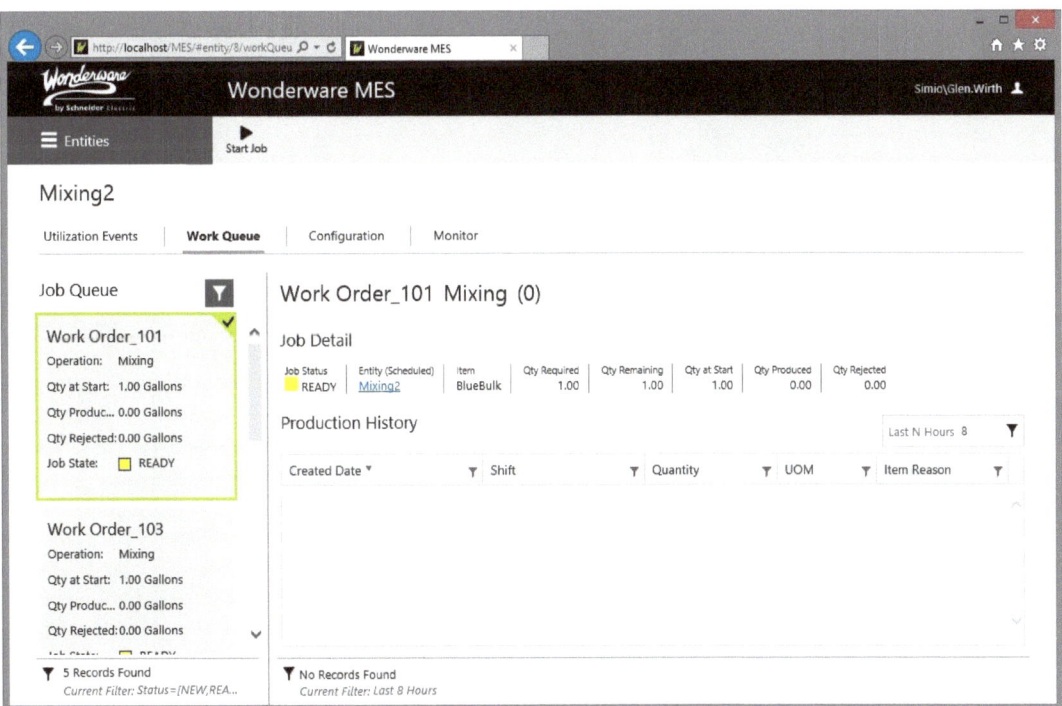

Appendix B – Publishing the Schedule using the Simio Portal

An alternative method for publishing the Simio schedule uses the Simio Portal. The Simio Portal is a web based model repository used to publish the production schedule to be viewed throughout the organization. The Simio Portal can be hosted within an enterprise or hosted on Microsoft Azure.

In this example, we will publish the Simio schedule to Microsoft Azure. Once the user logs into the Simio portal, they will be provided a number of capabilities based on their permissions. This user has been given full access in the portal.

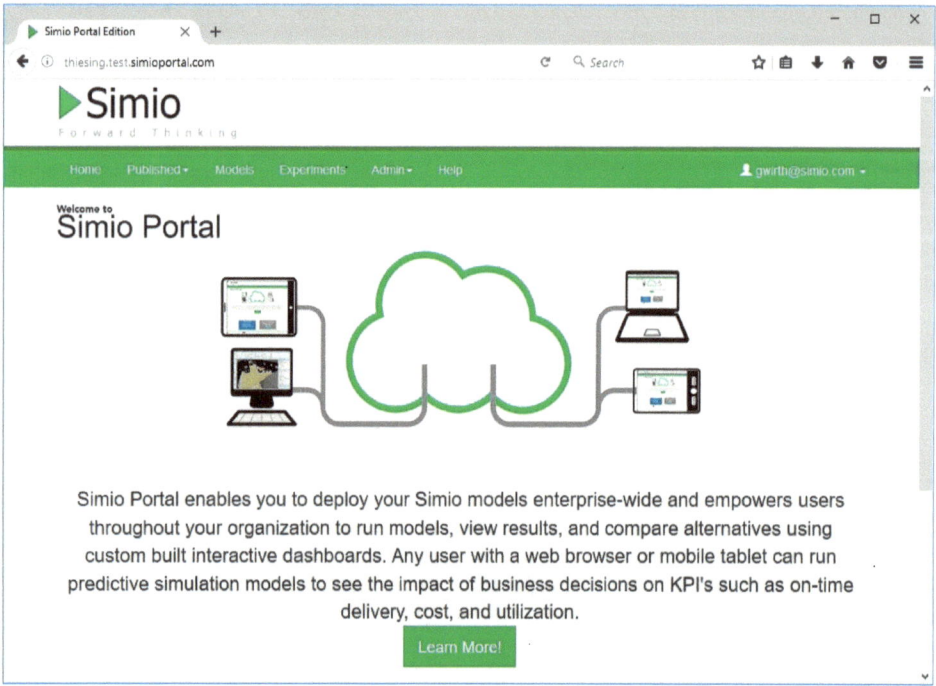

The user selects Published to upload a model.

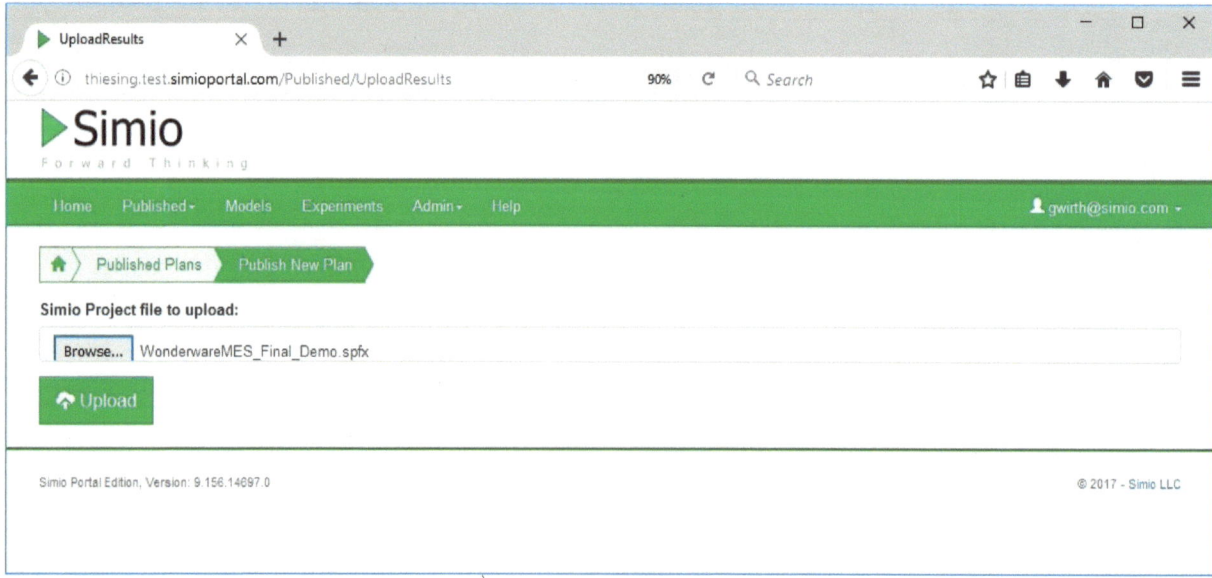

Once the model is uploaded to the Portal, the user will have the option to display Dashboards, Table Reports, Results, Gantt Charts and Logs.

Under Gantt Charts, select 'By Resource'. This will display the Resource Gantt chart

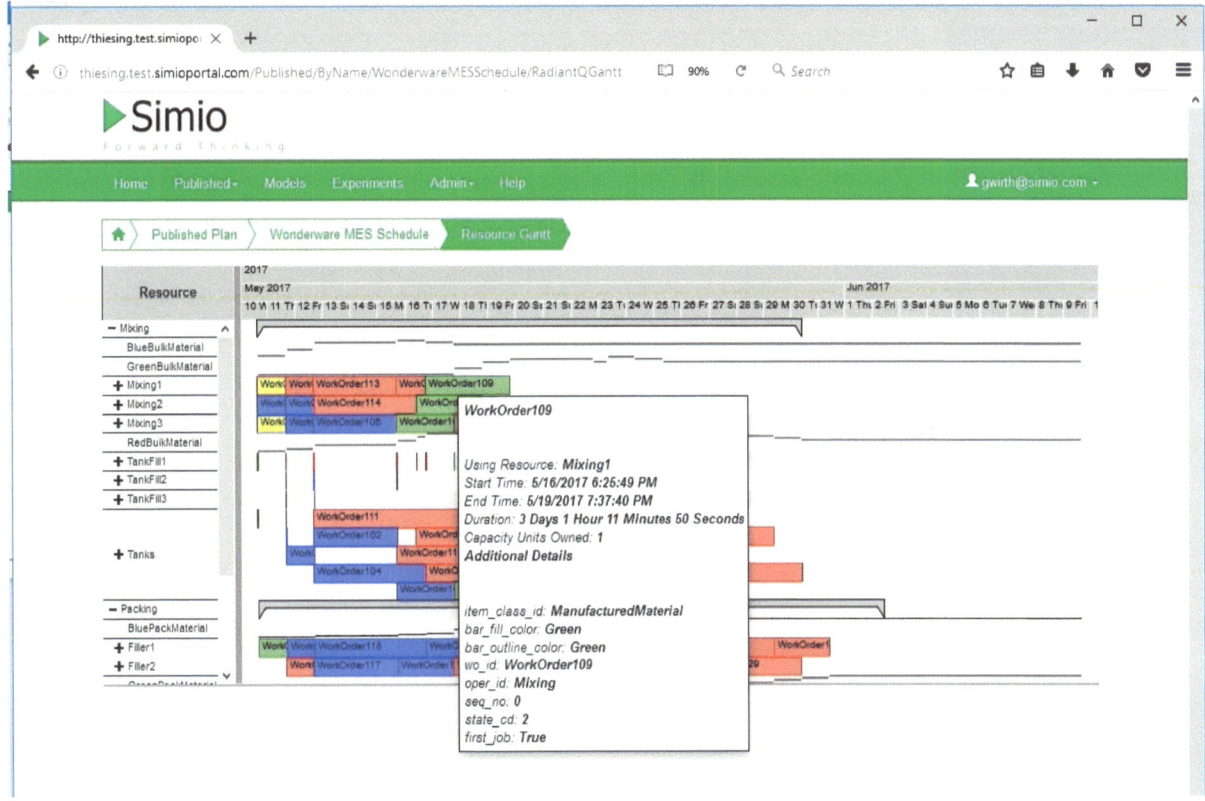

Dashboards and Reports can be configured so only certain users can view them.

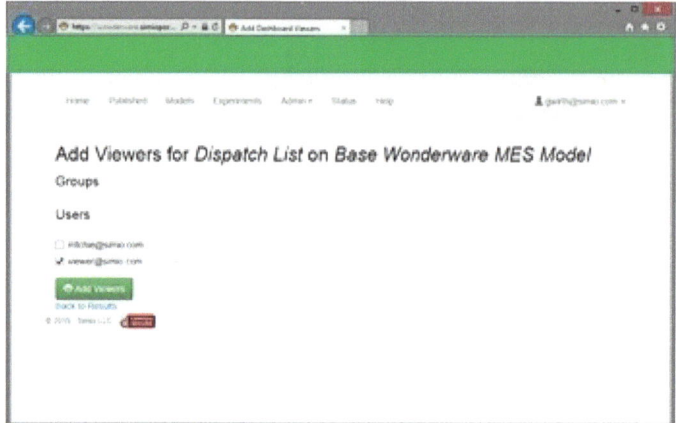

When the viewer@simio.com user access the schedule, they only view the Dispatch List

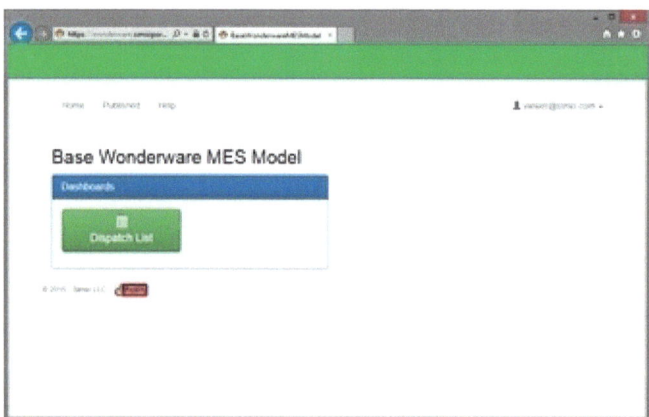

Here is an example of the Dispatch List Dashboard provided to the viewer@simio.com user.

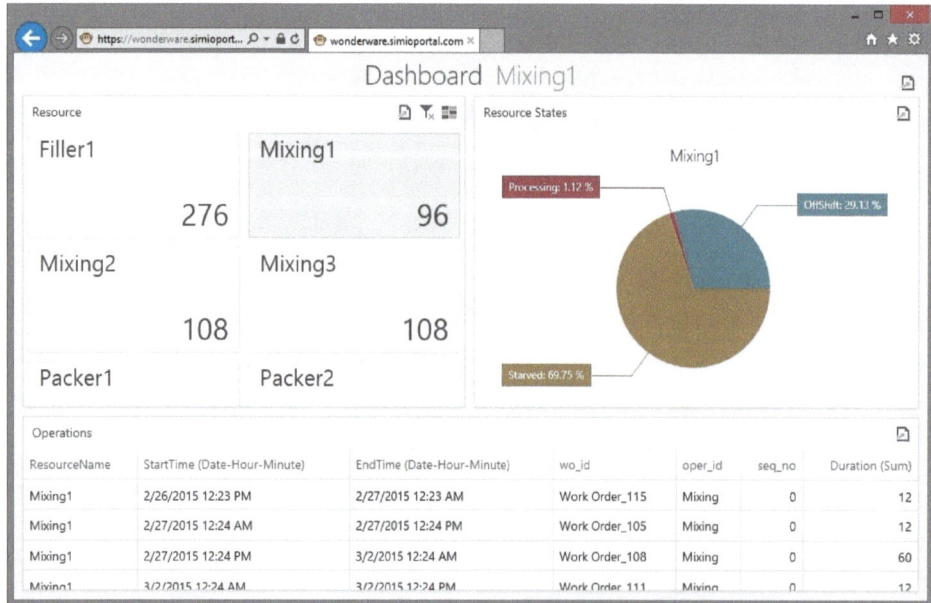